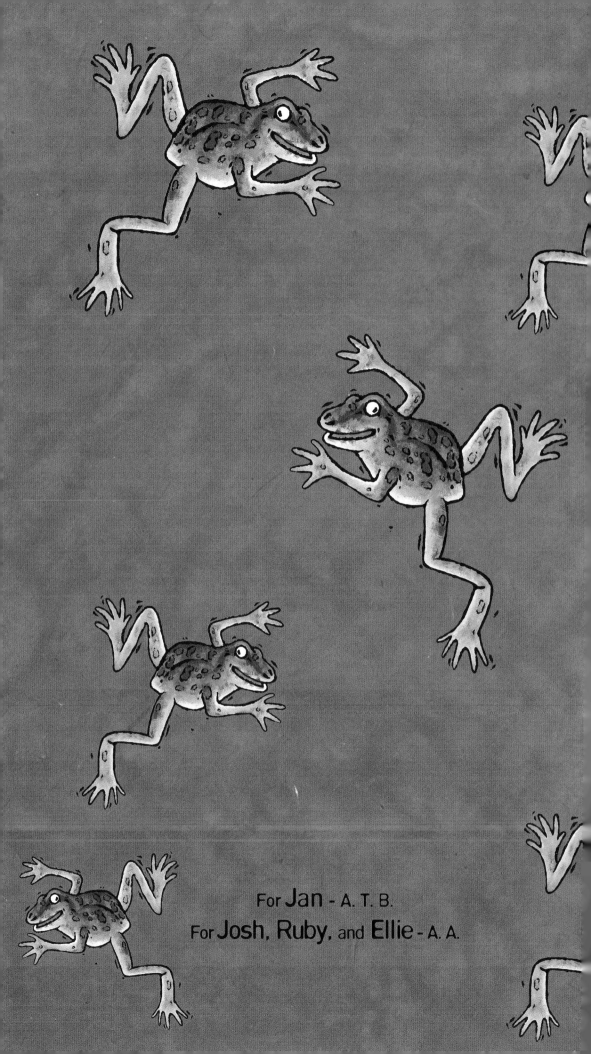

For Jan - A. T. B.
For Josh, Ruby, and Ellie - A. A.

Let's
Leap
and
Jump

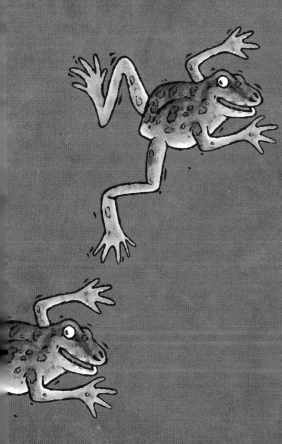

Please visit our web site at: www.garethstevens.com
For a free color catalog describing Gareth Stevens' list of high-quality books and
multimedia programs, call 1-800-542-2595 (USA) or 1-800-461-9120 (Canada).
Gareth Stevens Publishing's Fax: (414) 332-3567.

Library of Congress Cataloging-in-Publication Data

Nilsen, Anna, 1948-
　　Let's leap and jump / written by Anna Nilsen; illustrated by Anni Axworthy.
　　　　p. cm. – (Animal antics)
　　ISBN 0-8368-2912-3 (lib. bdg.)
　　1. Animals–Miscellanea–Juvenile literature.　2. Animal jumping–Juvenile literature.
　[1. Animal jumping.　2. Jumping.]　I. Axworthy, Anni, ill.　II. Title.
　QL49.N542　2001
　590–dc21　　　　　　　　　　　　　　　　　　　　　　　　2001020883

This North American edition first published in 2001 by
Gareth Stevens Publishing
A World Almanac Education Group Company
330 West Olive Street, Suite 100
Milwaukee, WI 53212 USA

Gareth Stevens editor: Dorothy L. Gibbs
Cover design: Tammy Gruenewald

This edition © 2001 by Gareth Stevens, Inc. First published by Zero to Ten Limited, a member of the
Evans Publishing Group, 327 High Street, Slough, Berkshire SL1 1TX, United Kingdom. © 1999 by Zero
to Ten Ltd. Text © 1999 by Anna Nilsen. Illustrations © 1999 by Anni Axworthy. This U.S. edition
published under license from Zero to Ten Limited.

Printed in the United States of America

1 2 3 4 5 6 7 8 9 05 04 03 02 01

Let's Leap and Jump

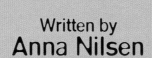

Written by
Anna Nilsen

Illustrated by
Anni Axworthy

Gareth Stevens Publishing
A WORLD ALMANAC EDUCATION GROUP COMPANY

Climbing trees
to gather nuts,
busy squirrels

leap
and
jump.

For quick
escapes, frogs
can go sideways
when they

leap
and
jump.

Traveling through treetops,

gibbons gibber-jabber as they

leap

and

jump.

Hunting cats
wait quietly
before they

leap
and
jump.

Swift impalas
bounce and
glide as they

leap
and
jump.

Look out,
salmon!

Grizzly bears
are watching
when you

leap

and

jump.

Kangaroos
have strong
back legs
to help them

leap
and
jump.

Tiny fleas
spring sky-high
when they

leap
and
jump.

Even boys and
girls like you
sometimes

leap
and
jump.